WHY OUR HORSES ARE BAREFOOT

*Everything We've Learned About
the Health and Happiness of the Hoof*

Published in the United States by 14 Hands Press,

an imprint of Camp Horse Camp, LLC

www.14handspress.com

Some of the material herein has appeared

in other books by Joe Camp

Library of Congress subject headings

Camp, Joe

Why Our Horses Are Barefoot / by Joe Camp

Horses

Human-animal relationships

Horses-health, hoof care

Horsemanship

The Soul of a Horse: Life Lessons from the Herd

ISBN 978-1-930681-41-5

First Edition

WHY OUR HORSES ARE BAREFOOT

*Everything We've Learned About the
Health and Happiness of the Hoof*

JOE CAMP

14 HANDS PRESS

Also by Joe Camp

The National Best Seller
The Soul of a Horse
Life Lessons from the Herd

The Soul of a Horse Blogged
The Journey Continues

Who Needs Hollywood
The Amazing Story of a Small Time Filmmaker
Who Writes the Screenplay, Raises the Production Budget,
Directs, and Distributes the #3 Movie of the Year

The Benji Method
Teach Your Dog to do what
Benji Does in the Movies

Horses Were Born To Be On Grass
How We Discovered the Simple But
Undeniable Truth About Grass, Sugar,
Equine Diet & Lifestyle

Horses Without Grass
How We Kept Six Horses Moving and Eating
Happily Healthily on an Acre and a Half
of Rocks and Dirt

Beginning Ground Work
Everything We've Learned About
Relationship and Leadership

Why Relationship First Works
Why and How It Changes Everything

Training with Treats
With Relationship and Basic Training
Locked In Treats Can Be an Excellent Way
to Enhance Good Communication

"Joe Camp is a master storyteller." - *THE NEW YORK TIMES*

"Joe Camp is a natural when it comes to understanding how animals tick and a genius at telling us their story. His books are must-reads for those who love animals of any species." - *MONTY ROBERTS, AUTHOR OF NEW YORK TIMES BEST-SELLER THE MAN WHO LISTENS TO HORSES*

"Camp has become something of a master at telling us what can be learned from animals, in this case specifically horses, without making us realize we have been educated, and, that is, perhaps, the mark of a real teacher. The tightly written, simply designed, and powerfully drawn chapters often read like short stories that flow from the heart." - *JACK L. KENNEDY, THE JOPLIN INDEPENDENT*

"One cannot help but be touched by Camp's love and sympathy for animals and by his eloquence on the subject." - *MICHAEL KORDA, THE WASHINGTON POST*

"Joe Camp is a gifted storyteller and the results are magical. Joe entertains, educates and empowers, baring his own soul while articulating keystone principles of a modern revolution in horsemanship." - *RICK LAMB, AUTHOR AND TV/RADIO HOST "THE HORSE SHOW"*

Go to The Soul of a Horse Channel on YouTube
to watch Joe's video of Why Barefoot?

*For Kathleen, who loved me enough to initiate all this
when she was secretly terrified of horses*

CONTENTS

INTRODUCTION

Often, in the early evening, when the stresses of the day are weighing heavy, I pack it in and head out to the pasture. I'll sit on my favorite rock, or just stand, with my shoulders slumped, head down, and wait. It's never long before I feel the magical tickle of whiskers against my neck, or the elixir of warm breath across my ear, a restoring rub against my cheek. I have spoken their language and they have responded. And my problems have vanished. This book is written for everyone who has never experienced this miracle.

- Joe Camp
The Soul of a Horse
Life Lessons from the Herd

1

HALF WILD

Shortly after our first three horses had shown up in our front yard I stumbled into wild horse research trying to uncover why Cash had come to us with shoes on his front feet but none on his back feet. We had been told that concrete and asphalt would crack and shatter any horse's hoof that was not wearing a shoe. If that were true, I needed to get shoes on his back feet right now because we had concrete and asphalt everywhere.

And I suppose I had wondered – apparently too often too loud – how wild horses had managed to exist all this time without help from humans? Could it be the wild ones might be able to teach us a thing or two?

Oh no. Generations of selective breeding have completely changed everything about the domesticated horse.

I had been told that again and again. But then I happened upon Jaime Jackson's research on how wild horse hooves work and look, and why. And how well their hooves had protected them and helped them survive for millions of years.

To a prey animal like a wild horse there is nothing more important than good rock-solid feet. They travel up to thirty miles a day in search of food and water. And they run a lot from predators.

All without metal shoes.

But apparently I was wasting my time. That information was all immaterial. The wild horse and the domesticated horse were different species.

Everybody said so.

A few hundred years of selective breeding had made it so.

Domesticated horses no longer had the same feet as their wild counterparts.

A domesticated hoof was destined to be weak and underdeveloped. Often sick and unhealthy. A domesticated hoof needed a metal shoe.

The American Farriers Journal reported that 95% of all domesticated horses have some sort of lameness issue. That's why they have to wear shoes, I was told.

But my Cash came to us with only *two* shoes. On the front feet. Nothing on the rear feet.

So was he half wild and half domestic?

It was worrisome that his back end would be the wild part. The kicking end.

All of this was gnawing at the edges of logic.

There is no hoof lameness in the wild. And it runs rampant throughout the domestic scene.

But unlike everyone else we had encountered, Jaime Jackson believed that wild hoof mechanics were exactly the same as domesticated hoof mechanics, both depending upon the hoof to flex with every step taken. Like a toilet plunger. This flexing sucks an enormous amount of blood into the hoof capsule every time the hoof hits the ground; and pushes it back up the leg when the hoof leaves the ground. Like a mini-heart, pumping with every step. Among other things this keeps the hoof healthy and growing properly.

And what happens when a metal shoe is nailed to the hoof?

Nothing.

No flexing.

No blood.

No function.

The curtains were parting. A veil lifting. It wasn't about whether a horse was *domesticated* or *wild*. It was about blood circulation. And the effect of that circulation - or lack of it - on the health of the hoof.

No, no, no. We've unfortunately bred the foot right off the horse.

I smiled politely.

Knowledge is king.

I had just read in a scientific journal that it would take a minimum of 5000 years to change the base genetics of any species. Probably more, depending upon the circumstances. A few hundred years of selective breeding could have no affect whatsoever on base genetics.

Which is why a newborn foal will still be on his feet less than an hour after being born - thinking, learning, eating - and in less than four hours be ready and able to move out with the herd to stay away from predators. Even if he's born in a stall.

The genetics haven't changed.

It's also why the stresses, illnesses, and vices caused by being confined to a stall can be solved by allowing horses to be out with each other 24/7. And it's why barefoot "domestic" horses living out with a herd and eating a proper diet from the ground are able to develop rock-solid hooves that have no use for metal shoes.

I called our vet.

"The shoes are coming off ," I said.

"Uhh… whose?"

"All of 'em."

Silence.

Then, "Why not try one at a time and see how it goes?"

"Nope. All six."

In a week it was done.

And not one of our horses ever looked back. I had never seen a horse smile until the day Cash's shoes were removed.

Scribbles, our paint, had hooves that were so sick that he had to grow a whole new foot, from hairline to the ground. It took eight months. But then he too was a happy camper.

Scribbles – eight months after shoes - California

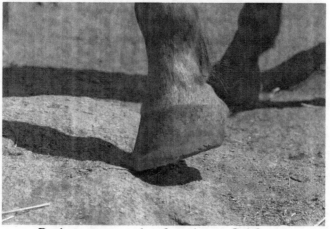

Pocket – six months after shoes– California

Cash – just a few weeks after shoes - California

How could this be? Domesticated horses need shoes because their feet are sick, soft, and unhealthy. Over and over again we were told. Or could it be that their feet are sick, soft, and unhealthy *because* of the metal shoes nailed to their feet restricting the circulation and eliminating the natural hydraulic-like shock absorption the intake of blood provides to protect the joints, ligaments, and tendons of the leg? I guess that's why the folks who run the country's largest mounted police patrol in Houston, Texas, have not one shoe on their forty or so horses who work all day every day on the concrete, asphalt, and marble of downtown Houston. Forty or so horses of every variety and every type of background. And not one shoe. Which eliminates the oft heard argument that *some* horses can go barefoot, but *some* cannot. All of theirs are barefoot and healthier for it. Unfortunately I didn't know any of this at the time.

"Which means that wild and domestic horses are not two different species at all," Kathleen said. "They're the same."

"I could turn Cash out into the wild and he'd be fine," I said.

"Down boy. That's taking research a step too far." She wasn't smiling.

The American Association for the Advancement of Science says that every horse on the planet "retains the ability to revert to living in the wild successfully." Note they use the word *successfully*.

DNA sequences taken from long bone remains of horses found preserved in the Alaskan permafrost dated 12,000 to 28,000 years ago differ by as little as 1.2% from the modern domestic horse.

So genetically speaking there was really no difference between a horse living in the wild and a so-called domesticated horse. What each horse has learned from his or her environment is obviously different, as I would soon come to understand. But their genetic ability to live successfully in the wild is the same.

And there was good news in the realization.

We had in our care perhaps the only species on the planet that lives with humans but could boast its own living laboratory in the wild. No more need for guesswork. These wild horses could reveal the truth, be a road map to the way horses were designed to live. A way that works because they have survived for 52 million years without any help from us.

If it hadn't worked so well we would've never heard of the horse.

God and Mother Nature knew what they were doing. Horses were designed over time through trial and error to live and eat and move in certain ways; and the study of all of this could provide more incredibly valuable information about how we should be feeding, keeping, and caring for the horses we choose to associate with than has ever been understood before.

2

MOTIVATION

How did we get into all this? Our heads were swimming.

And we had no idea where it was all leading. *Just tell us please how one is supposed to properly take care of a horse. Or six.* What did they need? How were they supposed to live? How should we approach them?

The pivotal point was the day my Cash arrived. I had discovered the wild horse studies of Monty Roberts. How and why horses communicate, socialize, and trust each other. So I took Cash immediately into the round pen and, using Monty's Join-Up techniques and language I gave Cash the choice of whether or not he wanted to trust me, to be in relationship with me. Many owners and trainers deny the horse that choice, forcing him into submission. But Monty's concepts are based upon letting the horse make the decision, on his own. And when he does, everything changes.

It was a scary moment because rejection is not one of my favorite concepts. What would I to do if Cash said *no thanks?* Mercifully the question didn't come up. And when he walked up behind me and touched me on the shoulder saying *I trust you to be my leader* it did, in fact, change everything. Cash has never stopped trying, never stopped giving. And I was no longer a horse owner. I was a friend, a companion, a partner, a leader. And I vowed at that moment, promised him out loud, I would provide for him the very best life that I possibly could.

The problem was: we didn't know what that was. We didn't know how to evaluate *the very best*. That's when I began asking questions. *Lots* of questions. And slowly we began to realize that many of the answers we were getting simply didn't make any sense.

So we began to dig around on our own and it wasn't long before Kathleen and I began to feel that we must be missing something because the manner in which most horses were being kept and cared for seemed very wrong when evaluated against the answers we were now finding.

We searched for that missing piece, for how could we rank neophytes who had no experience whatsoever be discovering things that the experienced and the qualified with decades of knowledge didn't know? It seemed like a silly notion. But as the puzzle pieces began to come together something else began to stir, like a baby chick pecking and poking its way out of an egg. I found myself toying with the idea of writing about this bizarre, funny, sad, amazing, and incredibly enlightening journey. To make an attempt to get some of these discoveries out amongst people who love their horses.

Joe and Cash cover photo – The Soul of a Horse

Kathleen said the title of the book should've been *Google It* because that's where I spent every waking moment when I wasn't out with the horses. Devouring studies, scientific papers, and archives. Which lead me into the middle of that herd of wild horses living free out in the great basin. It was there that the traditional thinking about horses cracked and crumbled with the discovery that equine genetics had been designed and developed over roughly fifty-two million years to help a prey animal survive in the wild. A prey animal who had no defenses whatsoever against predators except to flee. Genetics that were designed to live in wide open spaces where predators can be seen from long distances. Genetics that require living in a herd because there is safety in numbers. And genetics that require moving ten to thirty miles a day, all while eating small bits of mostly grass for up to eighteen to twenty hours a day to suit the tiny tummy and hindgut of this unique species.

Quite simply all of this blew me away. I suppose I had probably thought that horses were born with metal shoes nailed to their feet.

We were definitely in overload. I felt myself going *tharn*.

I love that word from *Watership Down*. It's rabbit-speak, and there is simply no English equivalent. It's what happens when a rabbit gets caught in the headlights and is so suddenly petrified that he can neither move nor think.

I was definitely heading toward a good case of tharn.

But I couldn't turn loose of my promise to Cash. And the illogical answers we continued to receive about how the horses were supposed to be cared for and why. All of that continued to haunt me.

Until I discovered that it would take a minimum of 5,000 years to even begin to change the base genetics of any species. Probably closer to 10,000 years. Which meant that there was no chance in the world that a few generations of selective breeding could do any harm to the genetics of the horse's hoof.

Who knew that every horse on the planet is genetically the same? A scientific truth. Why, I wondered, did I not know this? And does it make any difference? Quite bit, I discovered.

To the horse.

I have often wondered what might've happened if Cash had come to us with shoes on all four feet.

That first question might've never come up.

But more than that I have continuously wondered how it is that we have uncovered so much information that so few others seem to know about.

"Motivation," said Kathleen.

"Excuse me."

"You are a very emotional being."

"And?"

"When Cash made his own choice to trust you to be his leader, to be in relationship with you... well, it kind of did you in," she chuckled.

I was silent for a moment.

"Yup. It did. I was sniveling like a baby."

"From that moment you were determined to give him as much as he was giving you."

It was an uncomfortable moment. The eyes were beginning to mist.

"And you have," she said. "He has it all. He lives like a horse, and he has you."

"Us," I corrected.

"Like I said, you were motivated."

She paused, a tiny smile stretching across her face.

"And therein lies the very best thing about Monty's Join Up."

I thought about that for a moment.

"You got emotional, made a promise to Cash, and you've been obsessed ever since."

I smiled.

She was right. Thank you Monty.

3

WHY THE OUCH?

When a shoe comes off a horse that has been shod for years and years, the hoof and hoof wall are usually no longer strong and healthy. The hoof has been made unhealthy by lack of circulation because it has not been able to flex and thus circulate the blood properly throughout the hoof mechanism. And the continuing process of hammering nails into the hoof wall makes it weaker, and provides places (the nail holes) for chips and cracks to occur. Also, some hooves, if they're really in bad health, will be tender for a while after going barefoot. And the unknowing owner concludes that the tenderness means the horse *needs* shoes.

Not so.

The hoof will completely heal and remodel itself and grow strong new horn and hard calloused sole. This is a logical and normal process (see the Resources section at the back of the book, especially Pete Ramey's and Jaime Jackson's books and videos). It takes approximately eight months for a horse to grow a new hoof, from his hairline to the ground. If properly trimmed to mimic the way wild horses' hooves trim themselves during daily wear, a worst-case scenario for a horse to acquire a completely new, rock solid, healthy foot, then, is approximately eight months. Many horses are much quicker. As you read earlier, Cash was good to go from the first day his shoes came off. And a happy horse indeed. Four of our six never had a tender moment after going barefoot. One took four months to remodel, and one took almost eight months. And well worth the time.

But wait! When my horse's shoe falls off, he starts limping almost immediately. And when the shoe is nailed back on, suddenly he's fine. Doesn't hurt anymore. Proof that the shoe is better for him than barefoot.

Have you ever crossed your legs for so long that your foot goes to sleep? We all have, and we all know what's happening. The leg-cross has cut off the circulation to the foot, and with no circulation, the nerve endings lose their sensitivity and fail to work. The second you uncross, or stand up, the circulation returns, as do the nerve endings.

Ooohh! Ouch!

The same thing happens to a horse when a metal shoe is nailed on. The inability of the hoof to flex removes its ability to pump blood, virtually eliminating circulation in the hoof mechanism. Without proper circulation, the nerve endings quit transmitting, and the horse no longer feels the "ouch." When the shoe falls off, the circulation returns and suddenly he can feel again.

Whoa, what's that about??

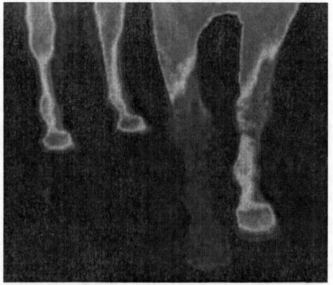

This is a thermograph of the blood circulation in the legs and hooves of a horse wearing one nailed-on metal shoe, on the front right. The other three hooves are barefoot. The rest is between you and your horse.

Convince yourself. Try this test. On a cool day, or under cover, feel the hoof and lower leg of a shod horse, and a hoof of another who is barefoot. The barefoot horse's lower leg and hoof will feel warm to the touch, because of the blood circulating within. The shod horse's lower leg and hoof will be cool, not warm, because of the blood that is not circulating within. Or simply study the thermograph above.

As mentioned earlier, our Scribbles took a good seven to eight months to regain a healthy hoof with no ouch. And today he's a happy camper, on asphalt or concrete, on the trail, in the arena. His hooves are beautifully concaved, keeping the coffin bone up where it belongs. They are beveled at the edges, just like a wild horse's hoof. And they're as hard as stone.

The sacrifice? The downside?

A few months time to let him grow the hoof nature always intended for him to have. Good boy, Scribbles.

How much trimming is needed, and how often, depends upon the horse's environment and lifestyle. Remember, the objective is to replicate the hoof that the horse would have if he were living in the wild, moving twelve to fifteen miles a day with the herd. If he's living in a Virginia pasture or, heaven forbid, a box stall and not moving around much, there will be a lot more trimming, probably more often, than if he were living in a high desert natural "pasture" like ours was in California and moving around all day wearing down his own hooves. But with a study of the next chapter the objective can be reached in any case.

Pete Ramey, a hoof specialist who teaches hoof care all over the world, believes we have only just begun to discover the true potential of the wild horse model. After a trip to wild horse country for research he said, "The country was solid rock; mostly baseball-sized porous *volcanic* rock that you could literally use as a rasp to work a hoof if you wanted to. Horse tracks were fairly rare, because there was so little dirt between the rocks.

There were a few muddy areas from the recent snow melt, but they were littered with rocks as well. The horses made no attempt to find these softer spots to walk on."

Pete and his wife, Ivy, observed, videotaped, and photographed at least sixty horses. All of them, from the foals to the aged, moved effortlessly and efficiently across the unbelievably harsh terrain. According to Pete, the horses were doing collected, extended trots across an obstacle course that would shame the best show ring work of any dressage horse, with their tails high in the air and their heads cocked over their shoulders watching the intruding couple.

"I have never known a horse I would attempt to ride in this terrain," Pete says. "Ivy and I had to literally watch our every step when we were walking. The movement of the horses was not affected by the slippery dusting of snow on the rocks. In fact, they got around much better than the mule deer and the pronghorns. The entire time we were there we did not see a limp, or even a 'give' to any rock, or a single lame horse, and not one chip or split in any of their hooves. It was an unbelievable sight."

The world has been shocked and amazed by the ability of Pete and others to forge rock-crushing bare hooves, boost equine performance, and treat "incurable" hoof disease. "I don't want to diminish these facts,"

Pete says, "but I now realize that we haven't even scratched the tip of the iceberg."

Pete is my hero! He is much of the reason for the existence of my book *The Soul of Horse*. He cares so very much about the horse and proves it daily by how much time he spends researching the complex internal workings of the horse's hoof and the myriad factors that affect its health. And substantiating that research on the horses in his care. In his DVD series, *Under the Horse*, Pete lays every bit of it out for us. If you've ever doubted the fact that horses do not need metal shoes and are in fact better off without them, please see this series. Pete will convince you of this undeniable truth. In my humble belief, whether or not you ever intend to trim your own horse's feet, *Under the Horse* is one of the most valuable and user-friendly compilations of knowledge, research, and insight for improving the health and lifespan of your horse that exists on the planet. And believe it or not, The American Farriers Association (the metal shoe group) agrees. An excerpt from their review:

"Pete Ramey's set of DVD's is without a doubt a must-have series for any equine professional farrier or horse owner. Pete Ramey offers wisdom and insight that his years of practice and study have given him. If you own only one hoof care DVD series, this should be it." *American Farriers Journal*

Pretty impressive for an organization of metal shoers.

There's an old expression: *No hoof, no horse.* And the reams of research I've pored over truly made that point. So much of what can go wrong with a horse begins or is controlled by the health of the hoof. When that hoof is healthy, is flexing, and taking stress off the heart, it can add years to a horse's life.

And he'll be happier.

Not only because he feels better, but because he can actually *feel* the surface he's walking on, which makes him more comfortable and more secure in his footing. As nature intended. Would you run on the beach with boots on? Or do you want to feel the sand between your toes? Not a perfect analogy, but you get the idea (See Healthier Happier Horses on our site).

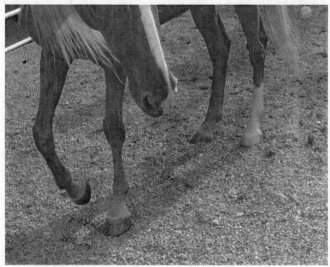

Mouse – a happy horse indeed – Tennessee

The president of the American Farrier's Association, in a speech to his constituency reported in the organization's publication, said that 95% of all the domestic horses on this planet have some degree of lameness. I wonder if he told them why. Or that it's a rare occurrence indeed to find a lame horse in the wild. Dr. Jay Kirkpatrick, director of the Science and Conservation Center in Billings, Montana, has studied wild horses most of his adult life and says that lameness in the wild is extremely rare and virtually every case he's seen is related to arthritic shoulder joints, not hoof problems.

All of the above is why our horses' shoes came off. How could I know all this, understand all this, and not do it. I had, after all, promised Cash.

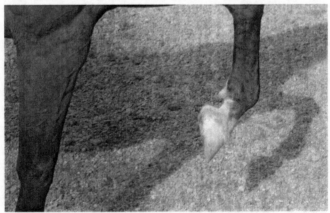

Cash - having fun - Tennessee

4

No Missing Pieces

Very quickly we learned that there was a bit more involved than just pulling four shoes.

Or twenty-four.

Quite a bit indeed.

Once barefoot with the wild horse model, I was in the soup, so to speak. I began to question and discover all sorts of other things. Like the importance of movement around the clock, a sugar-free diet, lifestyle with other horses, the horse's nature, the elimination of stalls, even blankets and leg wraps. And what I found was like a jigsaw puzzle. It's not just being barefoot that causes the wild horse to be healthier, happier, to be less stressed, to have better relationships. Take away one piece of the puzzle and it affects the whole picture. And the picture I was beginning to see more and more clearly was that we humans have completely manipulated horse care and training to suit ourselves, not the horse. Under what kind of leadership did we ever get to this place in time? Why has all this information not been front and center? I posed these questions to Dr. Matt, our vet, and began to think I was speaking to a politician up for election.

"Well, it's not always so black and white," was his answer to a question about going barefoot. "I like to see horses barefoot whenever they can be."

"Why not always?"

"Well, some horses have issues that others don't have."

"Like what, *owners?*"

The smile that wiggled across his lips betrayed his words.

"Well, some people feel their horses need shoes if they're going to be jumping, or, say, doing endurance riding."

"What do *you* think?" I asked.

"I think some horses have issues other horses don't have."

I was getting nowhere, and he had to keep moving. Another client to see. I knew from my own research and experience that he was an excellent vet. A caring vet who owned horses himself and loved horses. He had good communication with them. So I couldn't figure out why he was avoiding my questions. Was I wrong? Were there circumstances I hadn't yet stumbled upon in my digging that would refute the entire wild horse model? Was he simply not well informed on the work of this ever-growing band of natural trimmers and life-style designers?

I called the next day and asked if I could take him to lunch.

And told him why.

"I need to hear answers from you," I said. "I want to know what you really believe. I'll promise to never repeat what you say, if that's what you want, but I need to know that I'm not crazy. Everything I've been studying says that most of us are doing virtually nothing right in the way we care for our horses." What he told me the next day chilled my blood, and made me very sad. And I'm afraid it's only a microcosm of the way too much of our world works today.

A farrier is a person who makes a living putting metal shoes onto horses' natural feet by nailing into the horse's hoof wall. He used to be called a blacksmith. One organization reports that there are probably 100,000 farriers on the planet. The farrier's livelihood and self-esteem are generated by how well he appears to do his job. How well the shoe fits. How well it seems to solve some problem with the horse's foot, like an imbalance. Or an ouch. He decides whether the hoof needs a pad, or some packing, or wedges, or a special type of shoe. Often the farrier does no hoof trimming. His assistant does that. The farrier shapes the metal shoes and nails them on. One natural trimmer wrote about how difficult it was to stop shoeing and *just* be a trimmer using the wild horse model. He said it was an ego thing because his assistant usually did the trimming, and now he was more or less doing his assistant's work himself. He also called it a "male" thing, because he really liked the molding and shaping of steel, and hammering nails. I spoke with a couple farriers about switching to the concepts of trimming barefoot horses with the wild horse trim.

"It's bullshit," said one. "A horse needs shoes."

"Why?"

"Because we've bred the good foot right outta them."

"Have you ever tried it? The wild horse trim, I mean."

"Nope. Don't ever plan to."

"I know of several natural trimmers who have never been unsuccessful taking a horse barefoot."

"Anybody can take a horse barefoot. Just pull the shoes."

"I mean successful in that the horse never needed shoes afterward. Had a healthy, happy foot. On the trail. On the road. In the ring. Wherever."

"Bullshit," he said, which pretty much ended the conversation.

Not exactly my kind of logic.

There were a lot of horses in our community when we were in California, therefore a lot of farriers.

Dr. Matt told me how things are out in the field. He gets to see a client and treat a horse, usually, only when there's something wrong. An illness, or an injury. In other words, rarely. A farrier usually sees a client every six to eight weeks, maybe eight to twelve times a year. So most horse owners know their farrier much better than they know their vet. If it's a long-term relationship with the farrier, it would stand to reason that he is trusted. One bad word from the farrier about a particular vet, or a good word about some other vet, will be heard. And a farrier is not likely to recommend a vet who he knows is going to come in behind him and tell the owner to pull all the shoes off his horses.

Even with existing clients with whom he has had good relationships, Dr. Matt has lost patients because he recommended that shoes come off.

The owner calls the farrier about pulling the shoes.

The farrier explains that "most vets don't know much about feet because they don't work with feet. And, well, you should really think about it before pulling the shoes." Those words were actually spoken to me by a farrier.

In the above example, either the vet or the farrier is usually going to wind up losing a client because the last thing owners want are folks who disagree about the treatment of their horses. Especially if the owner doesn't have a clue about which one is right.

The very sad thing about all this is that all equine vets in the country should be educating themselves on the magical things that can be accomplished with the barefoot wild horse model. As should the universities housing their vet schools. And they should be talking to clients about it. But the truth is that it would be difficult indeed for a vet to make it in a community in which he has alienated all the farriers.

Cash – after a year in middle Tennessee

There's a vet in a neighboring community who actually stocks horseshoes and farrier tools and sells them to farriers at a discount. I'm guessing that doing so wins him a basket load of recommendations from farriers. Is he likely to tell his clients to pull off the shoes he himself sold to the farrier who nailed them on?

There is no disputing that a horseshoe prevents hoof flexing. Nor is there any dispute about why the hoof is supposed to flex. Nor about the good things that happen when it does. I can't help but wonder how a vet sworn to do his medical best for horses can sell horseshoes and supplies to farriers and still live with himself.

But I didn't press, and changed the subject with Dr. Matt.

"I've read that leg wraps are not good for horses," I said. "The article stated that they're usually tightly wrapped when the horse is at rest; then he goes out to work and the blood vessels in the leg attempt to dilate to get more blood down to the working leg, and the leg wrap prevents the vessels from dilating. True or false?"

Dr. Matt smiled. "I don't think they give any support or any true protection for the leg, but if they aren't worn too tightly, they don't really hurt anything."

"But, isn't it really best not to have them at all?" I persisted.

"Look at it this way, if an owner wants to use them, and I tell him no, and the horse comes up lame from some activity, who's going to be blamed?"

"Blankets?" I questioned, again changing the subject.

"No need for them unless it's really cold and raining. A horse has a terrific system for keeping his body temperature where it needs to be, unless his coat gets really soaked while it's really cold. Snow is no problem. It's the combination of cold and wet. I recommend pulling them as soon as the rain stops to keep the blanket from weakening the horse's own internal system."

"Is there any hard research on cold and wet?" I asked.

"Better safe than sorry," he said.

"So, most of the time, the owner is blanketing his horse to make himself feel better. Like I almost did."

"I prefer to think you were more misinformed than selfish." He smiled.

On the subject of stalls and barns, he did say that horses are better off moving around, being out 24/7. I was jubilant. At last an unqualified recommendation.

"Do you recommend that to your clients?" I asked.

"I try to be sensitive. If a person can do nothing but provide a twelve-by-twelve stall, there's little point in telling him to do something different."

I believe if a person loves his horse, he'll figure out a way to do what's best for him. Or at least put some thought into it. The more I've studied the more important movement, movement, movement has become. Horses are genetically structured to move eight to twenty miles a day. Their muscular, skeletal, and digestive structures depend upon it. But again, I didn't press.

"Sugar in the diet?" I questioned.

"The non-structured carbohydrate level in a horse's diet should not be above 10%."

Non-structured carbohydrates (NSC) are carbohydrates that turn into sugar immediately upon entering the body. Most processed "complete" feeds are well above that level.

"There's an old saying," Dr. Matt said. "You can have money. Or you can have horses. But you can't have both. I usually get called as the last resort because people don't want to spend money for a vet. Remember that call I got yesterday while I was at your place?

When I got over there, I was told the horse had been lying on his side without eating, pooping, or peeing for three days!

"*Three* days!" he added incredulously. "I deal with that kind of thing every day of my life."

"There are people who shouldn't have horses," I said.

He nodded.

I quietly thanked God for Dr. Matt, because I couldn't do what he does. I couldn't face what he faces every day. I'd have no clients at all by the end of the first week.

He did share that he felt there was a new day on the horizon. "For several generations, the horse was nothing more than a beast of burden, like an ox. Or a tractor."

"Or a motorcycle," I said.

"Right. But today I'm seeing more and more people who actually care for their horses. Granted, it's a small number, but it's growing. With all the publicity that people like you are getting, and the natural horsemanship clinicians, and the barefoot trimmers, and the vets who have studied all this... well, I have to believe it's getting better."

"I hope so," I said.

But, unfortunately, for the most part, word won't be coming from the folks whom you would normally turn to for advice. The farriers, most of them, are not going to take up the banner of barefoot. If they did, they would have to completely change their skill set or they'd be out of work. A few of them have done just that, but *few* is the operative word. A couple of my other books talk about how difficult it is to get most people to change, even when it's change for the better.

The veterinarians, most of them, have no choice but to hedge for the same reasons. Economics. Fear of being out of a job. Fear of risk. I've spoken to vets who have said, "I agree totally with what you're saying, but please don't tell anyone I said so." Many of them feel they can serve best by keeping their jobs and making a few small inroads here and there. Considering what they face, it's difficult to argue with that.

Until I look into my horse's eyes.

Letting him be a horse certainly won't be getting endorsements from the manufacturers of the horse shoes, forges, sugar-filled processed feeds, leg wraps, blankets, prefab barns, hay feeders, and so on. Those folks aren't going to burn their paychecks.

So think about it. Think seriously about it every time you hear someone say that what they do for a living is better for your horse than what the horse would do for itself in the wild. Ponder the presidents of those tobacco companies testifying before Congress, emphatic that tobacco was not harmful. Dig around on your own. Do some research. Compare what "the experts" say. Gather your own knowledge and don't let someone else make decisions for you. Whether it's about your horses, or your life.

And if you do own a horse, show him that he's not an ox, or a tractor, or a motorcycle. He's your partner.

And let him know by your leadership that you love him and will give him the best care you possibly can.

4

RECAP

The Undeniable Truths:

One: The first Undeniable Truth - Science tells us that it would take a minimum of 5000 years - probably closer to 10,000 – to even begin to change the base genetics of any species. In other words, no matter what anyone tells you to the contrary, a few hundred years of selective breeding has no effect on base genetics whatsoever or the horse's ability to grow the kind of rock solid foot he was born to have. This is the foundation for all that follows. Everything begins right here. The wild horse in the western high desert of the United States has incredible feet. He must have to escape predators and to search for food and water. If he didn't have incredible feet he'd be extinct. We would have never known him. And the wild horse and the domestic horse of today are genetically exactly the same. The domestic horse's foot is not genetically weak and unhealthy. Not even the oft-claimed Thoroughbred. The conditions under which any horse lives can certainly cause ill health, but the horse's genetics can fix that, given the opportunity.

Two: DNA sequencing was done on bones of horses discovered in the Alaskan permafrost dating 12,000 to 28,000 years old… and this DNA sequencing was compared to DNA sequencing from today's domestic horse… and there was less than 1.2% difference in those 28,000 year old horses and the horse in your back yard. Documented and on record. Confirming, once again, that the base genetics of every horse on the planet are the same. Science confirms for us that every horse on this earth "retains the ability to return successfully to the wild or feral state" – note that they say *successfully* – and that includes growing himself or herself a great foot that would protect this flight animal from predators and give him – or her – the ability to travel 8-20 miles every day of his life.

Three: The horse began and evolved for 50+ million years in and around the Great Basin of the western United States… then he crossed the Bering Straits Land Bridge into Siberia spreading into the rest of the world. Which means that the horse – as we know it today – spent 50+ million years evolving – now please get this because it's important – the horse spent 50+ million years evolving to live in conditions and on terrain like the western high desert of the United States and no horse will never adapt to the terrain and environment in our new home in middle Tennessee…or at least not for 5000 to 10,000 years… and it is therefore up to us – Kathleen and myself – to do everything within our power to replicate the lifestyle they would be living if they were living in the great basin – which is effectively the lifestyle they were living at our high desert home in southern California (See Horses Without Grass) before moving to middle Tennessee (See Horses Were Born To Be On Grass).

The herd in California

The herd in Tennessee

Five: Undeniable Truth #5 (or perhaps #1): a horse's hoof is supposed to flex with every impact of the ground. Every time it hits the ground it flexes outward – like a toilet plunger – and then snaps back when the hoof comes off the ground. That flexing sucks an enormous amount of blood into the hoof mechanism... keeps it healthy, helps it to grow properly, helps fight off problems... AND all that liquid provides an hydraulic-like shock absorption for the joints, ligaments, and tendons of the leg. Wow... who knew? At one point I remember believing the horse's hoof was just a wad of hard stuff... like one big fingernail.

But there's more. When the foot lifts off the ground and the flexed hoof snaps back, the power of that contraction shoves the blood in the hoof capsule back up those long skinny legs, taking strain off the heart.

So what happens to all this good stuff when a metal shoe is nailed to the hoof?

Nothing.

No circulation (or substantially reduced circulation)... no shock absorption (in fact if you've ever seen the videos of the vibrations set off up the leg when a metal shoe slams into the ground see link below, it'll freak you out)... and no assistance to the heart in getting that blood back up the leg.

Six: There is no hoof lameness in the wild. Yet the American Farriers Association reports that 95% of domestic horses have some degree of hoof lameness? Some folks want to say that's because the domestic hoof is inherently weak. But as we've already established, the inherent genetics are the same as the wild horse. The reasons for so much domestic hoof lameness are the metal shoes, diet, lifestyle, stress, and in some cases work load that we have forced upon the horse. In other words: No Stalls, no shoes, no sugar!

In simple terms, what all this means is that a horse's entire physiology has been built over millions of years to:

One: Move a minimum of 8 to 20 miles a day, <u>on bare hooves</u>.

Two: Be with a herd, and thus physically and emotionally safe, unstressed.

Three: Spend 16 to 18 hours a day eating... <u>from the ground</u>, a variety, but mostly grass or grass hay; a continuous uptake in small quantities to suit their small tummies and the function of their hindguts.

Four: Control their own thermoregulatory system, thus controlling their own internal body temperature with no outside assistance, including heat, blankets, and the like.

Five: Stand and walk on firm fresh ground, not in the chemical remnants of their own poop and pee... nor be breathing the fumes of those remnants, plus the excessive carbon dioxide that accumulates inside a closed structure. In other words, no stalls.

Six: Get a certain amount of unstressed REM sleep, which requires them to lie down, which will usually only happen when in the company of other horses, for guard duty.

My ten questions to ask a natural hoof specialist, a trimmer, <u>before</u> you hire him or her:

One: Are you exclusively barefoot? This will not set well with some trimmers but I believe strongly that if he or she does not so passionately believe in the history, the genetics, and the scientific facts of the wild horse trim and lifestyle that he or she would never nail a metal shoe onto a horse, then he or she is not a hoof specialist I would hire. When we moved to Tennessee I never got past this question with several that I interviewed. I just walked away. The best natural hoof professionals I know are passionate about being exclusively barefoot because they know a barefoot horse with proper diet and lifestyle will be a healthier, happier horse.

Two: Do you exclusively follow the wild horse model? Unfortunately there are a lot of folks who claim to be Natural Hoofcare Professionals who do not have a clue (or the wrong clue) why it works or what the wild horse model is all about. They might be okay, but probably not. Si I wouldn't risk my horses well being with one of these. I would pass them by.

Three: How important is diet and lifestyle to a successful barefoot experience? If they don't say (as do Eddie Drabek, and Pete Ramey, and so many others) that diet and lifestyle and <u>movement</u> are as important as the trim, then walk away.

Four: Do you incorporate the mustang roll? Eddie and Pete and others say this is the most important part of the trim. Must be used. Follow the wild horse example. If the answers to the above are all positive, then continue:

Five: May I have some references? Call clients of the hoof specialist and engage them in conversation. Get a sense.

Six: Where did you get your training? There are no right answers here, but very important. Still it's a judgment call, combined with all the other answers.

Seven: How long have you been natural trimming? Important, but not the end-all. But if only a short time it makes all the other answers even more important, especially the next two.

Eight: What sort of continuing education do you do? There is no right answer, but listen and be a good judge. If they say I don't really need any, walk away.

Nine: Who are your mentors or instructors you can go to when you need advice about a specific problem. Very important.

Ten: This one is not a question per se but very important: If the person gets irritable or defiant because of these questions, or if you feel like the answers are BS, avoid this person like the plague.

Bottom line: You must find a trimmer who knows and understands how to help the horse grow the foot his genetics know how to grow, not someone who wants to "cut" the foot the way he thinks it should look... and it must be someone both you and your horse feel comfortable with. Seriously. Listen to your horse on this.

Natural Hoof Specialist Eddie Drabek says, "I've had horses brought to me from owners who swore their horses had feet that grow abnormally, had bad genetics, could never be barefoot, have brittle hooves, heels grow but toes don't, toes grow but heels don't, have cracks that will never go away, and so forth... (I've heard it all)....but I've never met the horse who can't be taken successfully barefoot with the proper balanced trim and diet, and have beautiful feet to show for it. And I'm talking hundreds. I simply wouldn't have dropped everything and changed my entire career if I wasn't amazed with the results I was having. I was just at a horse show this past weekend watching many of my clients' horses compete. Not a little po-dunk show, rather a big time show with competitors from as far away as Canada and Australia. By popular belief about their high performance bloodlines and their 'genetically bad hooves', these horses should not be able to be barefoot, but they all are (most for 3 years now), competing right up there with the shod horses, and they have better feet than ever before."

Discovering the mysteries of the horse is a never-ending journey, but the rewards are an elixir. The soul prospers from sharing, caring, relating, and fulfilling.

Nothing can make you feel better than doing something good for another being. Not cars. Not houses. Not facelifts. Not blue ribbons or trophies. And there is nothing more important in life than love. Not money. Not status. Not winning.

Try it and you will understand what I mean. It is the synthesis of these books and why they came into being.

And please always question everything. Be your own expert. Gather information and make decisions based upon knowledge and wisdom, not hearsay. Know that if something doesn't seem logical, it probably isn't. If it doesn't make sense, it's probably not right.

When I gave Cash the choice of choice and he chose me, he left me with no alternative. No longer could it be what I wanted, but rather what he needed. What fifty-two million years of genetics demanded for his long, healthy, and happy life.

I could do it no other way.

Follow Joe & Kathleen's Journey
From no horses and no clue to stumbling through mistakes, fear, fascination and frustration on a collision course with the ultimate discovery that something was very wrong in the world of horses.

Read the National Best Seller
The Soul of a Horse
Life Lessons from the Herd

…and the new…

The Soul of a Horse Blogged
The Journey Continues

Go to
The Soul of a Horse Channel on YouTube
to watch the Video of Joe on Barefoot

The above links and all of the links that follow are live links in the eBook editions available at Amazon Kindle, Barnes & Noble Nook, and Apple iBooks

WHAT READERS AND CRITICS ARE SAYING ABOUT JOE CAMP

"Joe Camp is a master storyteller." *THE NEW YORK TIMES*

"Joe Camp is a gifted storyteller and the results are magical. Joe entertains, educates and empowers, baring his own soul while articulating keystone principles of a modern revolution in horsemanship." *RICK LAMB, AUTHOR AND TV/RADIO HOST "THE HORSE SHOW"*

"This book is fantastic. It has given me shivers, made me laugh and cry, and I just can't seem to put it down!" *CHERYL PANNIER, WHO RADIO AM 1040 DES MOINES*

"One cannot help but be touched by Camp's love and sympathy for animals and by his eloquence on the subject." *MICHAEL KORDA, THE WASHINGTON POST*

"Joe Camp is a natural when it comes to understanding how animals tick and a genius at telling us their story. His books are must-reads for those who love animals of any species." *MONTY ROBERTS, AUTHOR OF NEW YORK TIMES BEST-SELLER THE MAN WHO LISTENS TO HORSES*

"Camp has become something of a master at telling us what can be learned from animals, in this case specifically horses, without making us realize we have been educated, and, that is, perhaps, the mark of a real teacher. The tightly written, simply designed, and powerfully drawn chapters often read like short stories that flow from the heart." *JACK L. KENNEDY, THE JOPLIN INDEPENDENT*

"This book is absolutely fabulous! An amazing, amazing book. You're going to love it." *JANET PARSHALL'S AMERICA*

"Joe speaks a clear and simple truth that grabs hold of your heart." *YVONNE WELZ, EDITOR, THE HORSE'S HOOF MAGAZINE*

"I wish you could *hear* my excitement for Joe Camp's new book. It is unique, powerful, needed." *DR. MARTY BECKER, BEST-SELLING AUTHOR OF SEVERAL CHICKEN SOUP FOR THE SOUL BOOKS AND POPULAR VETERINARY CONTRIBUTOR TO ABC'S GOOD MORNING AMERICA*

"I got my book yesterday and hold Joe Camp responsible for my bloodshot eyes. I couldn't put it down and morning came early!!! Joe transports me into his words. I feel like I am right there sharing his experiences. And his love for not just horses, but all of God's critters pours out from every page." *RUTH SWANDER – READER*

"I love this book! It is so hard to put it down, but I also don't want to read it too fast. I don't want it to end! Every person who loves an animal must have this book. I can't wait for the next one !!!!!!!!!" *NINA BLACK REID – READER*

"I LOVED the book! I had it read in 2 days. I had to make myself put it down. Joe and Kathleen have brought so much light to how horses should be treated and cared for. Again, thank you!" *ANITA LARGE - READER*

"LOVE the new book… reading it was such an emotional journey. Joe Camp is a gifted writer." *MARYKAY THUL LONGACRE - READER*

"I was actually really sad, when I got to the last page, because I was looking forward to picking it up every night." SABINE REYNOSO - READER

"*The Soul of a Horse Blogged* is insightful, enlightening, emotionally charged, hilarious, packed with wonderfully candid photography, and is masterfully woven by a consummate storyteller. Wonderful reading!" HARRY H. MACDONALD - READER

"I simply love the way Joe Camp writes. He stirs my soul. This is a must read book for everyone." *DEBBIE K - READER*

"This book swept me away. From the first to last page I felt transported! It's clever, witty, inspiring and a very fast read. I was sad when I finished it because I wanted to read more!" *DEBBIE CHARTRAND – READER*

"This book is an amazing, touching insight into Joe and Kathleen's personal journey that has an even more intimate feel than Joe's first best seller." *KATHERINE BOWEN – READER*

Also by Joe Camp

Horses Were Born
To Be On Grass
*How We Discovered the Simple But
Undeniable Truth About Grass, Sugar,
Equine Diet & Lifestyle*

Horses Without Grass
*How We Kept Six Horses Moving and Eating
Happily Healthily on an Acre and a Half
of Rocks and Dirt*

Beginning Ground Work
*Everything We've Learned About
Relationship and Leadership*

Why Relationship First Works
Why and How It Changes Everything

Training with Treats
*With Relationship and Basic Training
Locked In Treats Can Be an Excellent Way
to Enhance Good Communication*

Who Needs Hollywood
*The Amazing Story of a Small Time Filmmaker
Who Writes the Screenplay, Raises the Production Budget,
Directs, and Distributes the #3 Movie of the Year*

The Benji Method
*Teach Your Dog to do what
Benji Does in the Movies*

RESOURCES

Taking Your Horse Barefoot: Taking your horses barefoot involves more than just pulling shoes. The new breed of natural hoof care practitioners have studied and rely completely on what they call the wild horse model, which replicates the trim that horses give to themselves in the wild through natural wear. The more the domesticated horse is out and about, moving constantly, the less trimming he or she will need. The more stall-bound the horse, the more trimming will be needed in order to keep the hooves healthy and in shape. Every horse is a candidate to live as nature intended. The object is to maintain their hooves as if they were in the wild, and that requires some study. Not a lot, but definitely some. I now consider myself capable of keeping my horses' hooves in shape. I don't do their regular trim, but I do perform interim touch-ups. The myth that domesticated horses *must* wear shoes has been proven to be pure hogwash. The fact that shoes degenerate the health of the hoof and the entire horse has not only been proven, but is also recognized even by those who nail shoes on horses. Successful high performance barefootedness with the wild horse trim can be accomplished for virtually every horse on the planet, and the process has even been proven to be a healing procedure for horses with laminitis and founder. On this subject, I beg you not to wait. Dive into the material below and give your horse a longer, healthier, happier life.

http://www.hoofrehab.com/– This is Pete Ramey's website. If you read only one book on this entire subject, read Pete's *Making Natural Hoof Care Work for You.* Or better yet, get his DVD series *Under the Horse*, which is fourteen-plus hours of terrific research, trimming, and information. He is my hero! He has had so much experience with making horses better. He cares so much about every horse that he helps. And all of this comes out in his writing and DVD series. If you've ever doubted the fact that horses do not need metal shoes and are in fact better off without them, please go to Pete's website. He will convince you otherwise. Then use his teachings to guide your horses' venture into barefootedness. He is never afraid or embarrassed to change his opinion on something as he learns more from his experiences. Pete's writings have also appeared in *Horse & Rider* and are on his website. He has taken all of Clinton Anderson's horses barefoot.

The following are other websites that contain good information regarding the barefoot subject:

http://www.TheHorsesHoof.com– this website and magazine of Yvonne and James Welz is devoted entirely to barefoot horses around the world and is surely the single largest resource for owners, trimmers, case histories, and virtually everything you would ever want to know about barefoot horses. With years and years of barefoot experience, Yvonne is an amazing resource. She can compare intelligently this method vs that and help you to understand all there is to know. And James is a super barefoot trimmer.

http://www.wholehorsetrim.com - This is the website of Eddie Drabek, another one of my heroes. Eddie is a wonderful trimmer in Houston, Texas, and an articulate and inspirational educator and spokesman for getting metal shoes off horses. Read everything he has written, including the pieces on all the horses whose lives he has saved by taking them barefoot.

Our current hoof specialist in Tennessee is Mark Taylor who works in Tennessee, Arkansas, Alabama, and Mississippi 662-224-4158 **http://www.barefoothorsetrimming.com/**

http://www.aanhcp.net- This is the website for the American Association of Natural Hoof Care Practioners.

Also see: **the video of Joe: Why Are Our Horses Barefoot? On** The Soul of a Horse Channel on YouTube.

Natural Boarding: Once your horses are barefoot, please begin to figure out how to keep them out around the clock, day and night, moving constantly, or at least having that option. It's really not as difficult as you might imagine, even if you only have access to a small piece of property. Every step your horse takes makes his hooves and his body healthier, his immune system better. And it really is not that difficult or expensive to figure it out. Much cheaper than barns and stalls.

Paddock Paradise: A Guide to Natural Horse Boarding This book by Jaime Jackson begins with a study of horses in the wild, then describes many plans for getting your horses out 24/7, in replication of the wild. The designs are all very specific, but by reading the entire book you begin to deduce what's really important and what's not so important, and why. We didn't follow any of his plans, but we have one pasture that's probably an acre and a half and two much smaller ones (photos on our website

www.thesoulofahorse.com). All of them function very well when combined with random
food placement. They keep our horses on the
move, as they would be in the wild. The big one
is very inexpensively electrically-fenced. *Paddock
Paradise* is available, as are all of Jaime's books,
at **http://www.paddockparadise.com/**

Also see the video **The Soul of a Horse Paddock Paradise: What We Did, How We Did It, and Why** on The
Soul of a Horse Channel on YouTube.

New resources are regularly updated on Kathleen's and
my: **www.theSoulofaHorse.com** or our blog
http://thesoulofahorse.com/blog

Liberated Horsemanship at:
http://www.liberatedhorsemanship.com/
Scroll down to the fifth Article in the column on the
right entitled Barefoot Police Horses

**An article about the Houston Mounted Patrol on our
website:** Houston Patrol Article

The next 2 links are to very short videos of a horse's
hoof hitting the ground. One is a shod hoof, one is
barefoot. Watch the vibrations roll up the leg from the
shod hoof... then imagine that happening every time

any shod hoof hits the ground: to view go to The Soul
of a Horse Channel on YouTube:

Video: Shod Hoof
Video: Barefoot Hoof

Find a recommended trimmer in your area:

http://www.aanhcp.net

http://www.americanhoofassociation.org

http://www.pacifichoofcare.org

http://www.liberatedhorsemanship.com/

<u>**Natural Horsemanship**</u>: This is the current buzz word
for those who train horses or teach humans the training
of horses without any use of fear, cruelty, threats, ag-
gression, or pain. The philosophy is growing like wild-
fire, and why shouldn't it? If you can accomplish everything you could ever hope for with your horse and still
have a terrific relationship with him or her, and be respected as a leader, not feared as a dominant predator,
why wouldn't you? As with any broadly based general
philosophy, there are many differing schools of thought
on what is important and what isn't, what works well
and what doesn't. Which of these works best for you, I
believe, depends a great deal on how you learn, and
how much reinforcement and structure you need. In
our beginnings, we more or less shuffled together

Monty Roberts and the next two whose websites are
listed below, favoring one source for this and another
for that. But beginning with Monty's Join-Up. Often,
this gave us an opportunity to see how different programs handle the same topic, which enriches insight.
But, ultimately, they all end up at the same place:
When you have a good relationship with your horse
that began with choice, when you are respected as your
horse's leader, when you truly care for your horse, then,
before too long, you will be able to figure out for yourself the best communication to evoke any particular objective. These programs, as written, or taped on DVD,
merely give you a structured format to follow that will
take you to that goal.

There are many programs and people who subscribe to
Natural Horsemanship philosophies and are very good
at what they do but are not listed in these resources.
That's because we haven't experienced them, and we
will only recommend to you programs that we believe,
from our own personal experience, to be good for the
horse and well worth the time and money.

Monty Roberts and Join up:
http://www.montyroberts.com- Please start here! Or at
Monty's Equus Online University which is terrific and
probably the best Equine learning value out there on
the internet (Watch the Join-Up lesson <u>and</u> the Special

Event lesson. Inspiring!). This is where you get the relationship right with your horse. Where you learn to give him the choice of whether or not to trust you. Where everything changes when he does. Please, do this. Learn Monty's Join-Up method, either from his Online University, his books, or DVDs. Watching his *Join-Up* DVD was probably our single most pivotal experience in our very short journey with horses. Even if you've owned your horse forever, go back to the beginning and execute a Join Up with your horse or horses. You'll find that when you unconditionally offer choice to your horse and he chooses you, everything changes. You become a member of the herd, and your horse's leader, and with that goes responsibility on his part as well as yours. Even if you don't own horses, it is absolutely fascinating to watch Monty put a saddle and a rider on a completely unbroken horse in less than thirty minutes (unedited!). We've also watched and used Monty's *Dually Training Halter* DVD and his *Load-Up trailering* DVD. And we loved his books: *The Man Who Listens to Horses, The Horses in My Life, From My Hands to Yours, and Shy Boy.* Monty is a very impressive man who cares a great deal for horses.

http://www.parelli.com- Pat and Linda Parelli have turned their teaching methods into a fully accredited college curriculum. We have four of their home DVD courses: *Level 1, Level 2, Level 3,* and *Liberty & Horse Behavior.* We recommend them all, but especially the first three. Often, they do run on, dragging out points much longer than perhaps necessary, but we've found, particularly in the early days, that knowledge gained through such saturation always bubbles up to present itself at the most opportune moments. In other words, it's good. Soak it up. It'll pay dividends later. Linda is a good instructor, especially in the first three programs, and Pat is one of the most amazing horsemen I've ever seen. His antics are inspirational for me. Not that I will ever duplicate any of them, but knowing that it's possible is very affirming. And watching him with a new-born foal is just fantastic. The difficulty for us with *Liberty & Horse Behavior* (besides its price) is on disk 5 whereon Linda consumes almost three hours to load an inconsistent horse into a trailer. Her belief is that the horse should *not* be *made* to do anything, he should *discover* it on his own. I believe there's another option. As Monty Roberts teaches, there is a big difference between *making* a horse do something and *leading* him through it, showing him that it's okay, that his trust in you is valid. Once you have joined up with him, and he trusts you, he is willing to take chances for you because of that trust, so long as you don't abuse the trust. On Monty's trailer-loading DVD Monty takes about one-tenth the time, and the horse (who was impossible to load before Monty) winds up loading himself from thirty feet away, happily, even playfully. And his trust in Monty has progressed as well, because he reached beyond his comfort zone and learned it was okay. His trust was confirmed. One thing the Parelli program stresses, in a way, is a followup to Monty Roberts' Join-Up: you should spend a lot of time just hanging out with your horse. In the stall, in the pasture, wherever. Quality time, so to speak. No agenda, just hanging out. Very much a relationship enhancer. And don't ever stomp straight over to your horse and slap on a halter. Wait. Let your horse come to you. It's that choice thing again, and Monty or Pat and Linda Parelli can teach you how it works.

http://www.downunderhorsemanship.com-
This is Clinton Anderson's site. Whereas the Parellis are very philosophically oriented, Clinton gets down to business with lots of detail and repetition. What exactly do I do to get my horse to back up? From the ground and from the saddle, he shows you precisely, over and over again. And when you're in the arena or round pen and forget whether he used his left hand or right hand, or whether his finger was pointing up or down, it's very easy to go straightaway to the answer on his DVDs. His programs are very task-oriented, and, again, there are a bunch of them. We have consumed his *Gaining Respect and Control on the Ground, Series I through III* and *Riding with Confidence, Series I through III.* All are multiple DVD sets, so there has been a lot of viewing and reviewing. For the most part, his tasks and the Parellis are

much the same, though usually approached very differently. Both have served a purpose for us. We also loved his *No Worries Tying DVD* for use with his Australian Tie Ring, which truly eliminates pull-back problems in minutes! And on this one he demonstrates terrific desensitizing techniques. Clinton is the only two-time winner of the Road to the Horse competition, in which three top natural-horsemanship clinicians are given unbroken horses and a mere three hours to be riding and performing specified tasks. Those DVDs are terrific! And Clinton's Australian accent is also fun to listen to… mate.

The three programs above have built our natural horsemanship foundation, and we are in their debt. The following are a few others you should probably check out, each featuring a highly respected clinician, and all well known for their care and concern for horses.

http://www.imagineahorse.com- This is Allen Pogue and Suzanne De Laurentis' site. I cannot recommend strongly enough that everyone who leaves this eBook Nugget ready to take the next step with treats and vocabulary should visit this site and start collecting Allen's DVDs (he also sells big red circus balls). Because of his liberty work with multiple horses Allen has sort of been cast as a trick trainer, but he's so much more than that. It's all about relationship and foundation. We are dumbfounded by how Allen's horses treat him and try for him. His work with newborn foals and young horses is so logical and powerful that you should study it even if you never intend to own a horse. Allen says, "With my young horses, by the time they are three years old they are so mentally mature that saddling and riding is absolutely undramatic." He has taken Dr. Robert M. Miller's book *Imprint Training of the Newborn Foal* to a new and exponential level.

http://www.chrislombard.com/ - An amazing horse-man and wonderful teacher. His DVD *Beginning with the Horse* puts relationship, leadership and trust into simple easy-to-understand terms.

Frederick Pignon – This man is amazing and has taken relationship and bond with his horses to an astounding new level. Go to this link:
(http://www.youtube.com/watch?v=w1YO3j-Zh3g)
and watch the video of his show with three beautiful black Lusitano stallions, all at liberty. This show would border on the miraculous if they were all geldings, but they're not. They're stallions. Most of us will never achieve the level of bond Frederick has achieved with his horses but it's inspiring to know that it's possible, and to see what the horse-human relationship is capable of becoming. Frederick believes in true partnership with his horses, he believes in making every training session

fun not work, he encourages the horses to offer their ideas, and he uses treats. When Kathleen read his book *Gallop to Freedom* her response to me was simply, "Can we just move in with them?"

http://www.robertmmiller.com - Dr. Robert M. Miller is an equine veterinarian and world renowned speaker and author on horse behavior and natural horsemanship. I think his name comes up more often in these circles than anyone else's. His first book, *Imprint Training of the Newborn Foal* is now a bible of the horse world. He's not really a trainer, per se, but a phenomenal resource on horse behavior. He will show you the route to "the bond." You must visit his website.

New resources are regularly updated on Kathleen's and Joe's: www.theSoulofaHorse.com or their blog http://thesoulofahorse.com/blog

The following are links to videos on various subjects, all found on our The Soul of a Horse Channel on YouTube:

Video of Joe: Why Are Our Horses Barefoot?

Video of Joe: Why Our Horses Eat from the Ground

Video: Finding The Soul of a Horse

Video of Joe and Cash: Relationship First!

Video: The Soul of a Horse Paddock Paradise: What We Did, How We Did It, and Why

Don't Ask for Patience – God Will Give You a Horse

Valuable Links on Diet and Nutrition:

Dr. Juliette Getty's website:
http://gettyequinenutrition.biz/

Dr. Getty's favorite feed/forage testing facility:
Equi-Analytical Labs:
http://www.equi-analytical.com

For more about pretty much anything in this book please visit one of these websites:

www.thesoulofahorse.com

http://thesoulofahorse.com/blog

www.14handspress.com

thesoulofahorseblogged.com

The Soul of a Horse Fan Page on Facebook

The Soul of a Horse Channel on YouTube

Joe and The Soul of a Horse on Twitter
@Joe_Camp

CPSIA information can be obtained at www.ICGtesting.com
Printed in the USA
LVOW13s1249191113

361932LV00001B/99/P